D1534387

LEARN CODING BASICS IN HOURS WITH PYTHON

™

An Introduction to Coding for People with No Prior Experience

**Written by: Jack C. Stanley and Erik D. Gross,
Co-Founders of The Tech Academy**

Table of Contents

PREFACE

The Tech Academy is a licensed career school that delivers software developer training. Our main offering is the Software Developer Boot Camp. The program can be taken online from anywhere or at one of our campuses.

Please keep in mind that this is the second edition of this book. There will be upgrades and revisions in the future.

If you find any errors, have feedback or would like to give suggestions, please contact us immediately. Our contact information can be found on our website: learncodinganywhere.com.

Thank you for obtaining this book. We hope that you enjoy it!

INTRODUCTION

Welcome!

This book will be useful for individuals who are interested in learning coding and the basics of Python. Don't worry if you haven't ever coded before, though; this book doesn't require any previous knowledge or experience.

If you have written code before, this book isn't recommended because it was made for total beginners (people with absolutely *no* prior experience).

This book should be read start to finish, with all assignments completed in order. All tasks and code you are required to write herein are written in gray font. As a developer-in-training, you need all the experience you can get in typing code! So, definitely type all of the code in this book.

As this book is meant for newcomers, we won't start writing code until Section Three. The first couple sections, we will be covering programming fundamentals and the required vocabulary. Please be patient because writing code without any understanding of the basics of coding is a waste of time and creates nothing more than a parrot.

The information presented in this book is arranged in a gradient sequence—we will start with the easiest information first and end with the most difficult tasks.

The purpose of this book is to create: *A programmer who understands the basics of Python.*

This book was written by The Tech Academy. We are a licensed career school that trains students in computer programming and web development. If you're interested in finding out more about us, please visit our website: https://www.learncodinganywhere.com/.

Let's begin this book with some fundamental terms to orient you to the subject.

CHAPTER ONE

The product of this chapter is a student that understands the basic terminology associated with programming and Python.

Understanding the terminology of a subject is a vital component to mastering it. After all, how well could you drive a car if you didn't know what a gas pedal was?

So, while it may be boring, read through the following technological definitions so we can ensure a firm foundation is in place.

Note: If there are any technical terms you would like to clear up that aren't defined in this book, you can refer to The Tech Academy's *Technology Basics Dictionary, Tech and Computers Simplified*. The dictionary is available for purchase on Amazon.

Definitions

Machine: Something that uses energy to do things. Machines are most commonly made out of metal and plastic and have several parts that work together. They are usually made up of fixed and moving parts that each serve a particular purpose. Machines are used by people as tools to assist in getting something done.

EXAMPLE: A car is a machine.

Computer: An electronic machine that stores and deals with information. A computer is made up of many different parts through which electricity passes. Computers are a tool that people use to help them do things.

Computers follow instructions entered into them by people. A computer can store instructions to be done at later points in time so that people do not have to re-enter the same instructions over and over again. The computer automatically performs a series of functions when it is on and also responds to commands that a person enters into it. The way it responds to commands is all set up beforehand by people.

Machines, such as cars and phones, can have computers inside them that perform specific functions. Computers typically exist to help people by performing repetitive activities, storing information and making certain activities faster or more efficient.

EXAMPLE: An electronic calculator is a very basic type of computer.

Digital: Of, or related to, a device or object that represents magnitudes in digits. Computers are digital devices; data in computers is represented using digits.

The opposite of "digital" can be considered to be "analog." Analog refers to data flowing out in a continuous stream.

Things that are digital are made up of exact, distinct parts; these parts are always in an exact, standard state (state is the condition of a thing—for example, "green," "empty," "45 years old," etc.). Digital things can only be in one of the available states for that thing and not "in between" two of the available states. For example, a light bulb with a regular flip switch would be digital because it has only two states: totally off or totally on. If the light bulb had one of those round dimmer switches instead of a flip switch, and it could be set to somewhere between "totally off" and "totally on," it would not be digital.

A black-and-white photo that is created using only small black and white dots is digital. From a distance, you see the image that the photograph represents. If you look at it up close, you see that the photo is composed of individual small dots. The distinct dots are digital in that they are exact, separated marks.

EXAMPLE: In terms of physical universe objects, a staircase would be considered digital, and a ramp would be analog. Stairs have distinct, separated steps. A ramp is continuous.

Computers are digital because they communicate using two numbers: 0 (which tells the computer "off") and 1 (which tells the computer "on").

Note: People typically use the word "digital" to refer to anything that has to do with computers.

Central Processing Unit: "CPU" for short. The CPU is basically the "brain" of the computer.

A Central Processing Unit is the part of a computer that controls all the actions the computer does. It is a small device that contains billions of tiny electronic components.

Most CPUs have a set of built-in "instructions" in them. They define the various actions the CPU can take—things like taking in data from a keyboard, storing data in the computer, sending data to a display, etc.

The basic job of all processors is to execute various combinations of those simple instructions built into them. It only ever performs one of these instructions at a time, but modern computers can perform billions of instructions per second.

All actions you can make a computer do will be composed of various combinations of these built-in actions

EXAMPLE: If you have a list of students in a computer file, and you give the computer an instruction to put the list in alphabetical order, it is the CPU that performs the analysis and re-ordering of the list.

Peripheral: These are things that can be hooked up to a computer, including external devices. Peripherals include mice, keyboards, external drives, displays, etc. Peripherals are often used to get data into a computer or to receive data from the computer.

EXAMPLE: A printer is a peripheral device.

Input: Input is data or information that is collected by a computer. This can take many forms. It may be information a user has typed into a form on the computer. It may be electronic signals sent to the computer by an attached device, like a mouse or a display screen that takes touch input. It may be a set of electronic data from another computer that is connected to the first one by wires.

EXAMPLE: A customer service agent at a retail store takes down your information and types it into her computer; that information is input for the computer to use.

Algorithm: A mathematics word that means a plan for solving a problem. Algorithms consist of a sequence of steps to solve a problem or perform an action. Computers use algorithms. An algorithm is a set of instructions that is used to get something done.

EXAMPLE: An algorithm for selecting the right kind of shirt might have the following steps:

1. Pick a shirt from your closet.
2. Put the shirt on.
3. Look at yourself in the mirror.
4. Decide whether you like the way you look in that shirt. If you like how you look, leave the shirt on and go to step 6. If you do not like how you look, take the shirt off and put it back in the closet where you got it from.
5. Repeat steps 1–5 until you have made a decision about each shirt in the

closet.

6. End this procedure.

Instruction: A command or set of commands entered into a computer that performs a certain operation. Instructions control a computer and tell it what to do.

EXAMPLE: You could make a computer draw a square by typing in instructions.

Program: Programs are written instructions entered into the computer that make it perform certain tasks.

Programs are the things on a computer that you interact with to get things done. You use a program of one sort or another for basically everything you do on your computer, whether that is checking the weather or looking up banking information. There are countless programs that people have created over the years to do countless different things.

EXAMPLE: "Paint" is a program on many computers that allows users to draw and color things.

Programmer: A computer programmer is someone who can create computer programs. Another term for programmer is developer.

A programmer can type exact commands and instructions in a computer to create many of the things we use computers for on a day-to-day basis.

EXAMPLE: Microsoft Word (a program mainly used to type documents) was created by programmers.

Programming Language: In order to understand the term "programming language," you will need to understand the term "language."

Languages are communication systems that allow you to transfer ideas in written and spoken words.

Programming languages are organized systems of words, phrases, and symbols that let you create programs.

Programming languages are also called computer languages.

There are many different types of programming languages, each of which was created to fill a specific purpose. Usually, a language is created in order to make the creation of certain types of computer programs easier.

EXAMPLE: Python is a popular computer language. This is how you would tell a computer to display the words "Hello, world!" on the screen using the computer language Python:

print ("Hello, world!")

Machine Language: A machine instruction is an instruction to your computer written in a form the CPU can understand—machine code or machine language.

Machine language is data that is composed of a series of 1s and 0s; computers are operated using 1s and 0s since it is very easy to make a machine that only has to differentiate between two different states.

When computers are manufactured, the CPU is made so it will obey certain instructions and perform exact actions when given those instructions. The machine instructions are available from the manufacturer of the computer.

EXAMPLE: Let's say "10010100 01101011 01011011 01010110" told the computer to "Delete the file called Vacation.doc." The "10010100 01101011 01011011 01010110" is a machine instruction.

It is difficult for people to read and write in machine language since the instructions are just unique patterns of ones and zeroes.

Low-level language: A low-level language is a programming language that is at the same level as or just above machine language in terms of readability—that is, it uses codes to represent specific machine language instructions or different combinations of machine language instructions.

These codes are not the ones and zeros of machine language; rather, they might look like this:

MOV AL, 61h

This might be codes that mean "store the amount 97 in a special location in the CPU for later use."

This is more readable than ones and zeros of machine language but is still not very easy to work in.

Code: There are a couple of definitions for "code." The first is: what you type into a computer to make programs. Code is written in specialized computer languages. Coding means typing out instructions (using a particular computer language) to make a program that will make the computer perform certain actions.

EXAMPLE: You can write code that makes a computer game.

Code can also mean a system where you use a symbol to represent another thing.

EXAMPLE: Airports are often referred to by three-letter codes—the Portland International Airport in Portland, Oregon, USA has the three-letter code "PDX." So here, PDX is a code that represents the Portland International Airport.

Operating System: Abbreviated OS. An operating system is a special-purpose computer program that supports the computer's basic functions, such as scheduling tasks, running other computer programs, and controlling peripherals (external devices such as keyboards, mice, and displays).

Most computer programs will require that an operating system already be installed on a computer before they can function on that computer.

Nearly all computers available today come with an operating system already installed when they are purchased. Computer manufacturers install the operating system before they sell a computer.

Some operating systems are free; others are available for a fee.

One of the most well-known and popular operating systems in the world is called Windows. It is created and sold by the technology company Microsoft.

Other popular operating systems are: OS X (created by the technology company Apple; it is used on their desktop computers and laptops), Linux (a family of free and for-fee operating systems; it is used on desktop computers and laptops), Android (owned by the technology company Google; it is used on mobile devices like smartphones), and iOS (created by Apple; it is used on their mobile devices like the iPhone and iPad).

EXAMPLE: Windows 10 is a popular operating system.

Network: A network is a system where two or more computers are connected to each other. The computers can be connected by a cable (wired) or connected wirelessly. Network is the word used to describe a link between things that are working together. Networks are used in many different ways with computers.

EXAMPLE: Information can be shared from computer to computer through the use of a network.

Internet: The Internet is two things: an interconnected network of many computers around the world and a set of methods or protocols for transferring different types of data between those computers.

This is the basic definition of the Internet. A detailed description follows.

A protocol is a formal description of how a certain type of information will be formatted and handled. Basically, it's an agreement that the various people who work with that type of information all adhere to. Protocols are usually described in written documents and are very precise. They are usually created by experts in the applicable industry.

An example of a type of data where a protocol would be valuable is healthcare information. If various organizations in the healthcare industry were to transfer healthcare data back and forth between computers as they perform their work, it would be important that they all agree about things like the exact format of the information, how to keep private data safe, etc. All of that would be laid down in a written protocol. Several such protocols do exist in the healthcare industry.

The Internet is a network of connected computers, and there are lots of different types of data that can be sent back and forth between these computers—things like electronic messages, electronic documents, healthcare records, etc. One or more protocols have been created for each type of data that can be transferred around on the Internet.

EXAMPLE: A bank could devise a protocol for how to format and exchange financial data between its headquarters and its various branches and then use their computers connected to the Internet to actually exchange that data.

World Wide Web: Abbreviated WWW; usually referred to as "the Web." To understand what the World Wide Web is, you need to know about these other terms: website, web page, web browser, and web server.

The World Wide Web (the "Web") is a collection of linked electronic documents organized into groups called websites.

The Web is accessed by connecting to the Internet. The computers involved in the operation of the Web are connected to the Internet.

A website is composed of one or more individual web pages, where a "page" is an organized, separate document containing text, images, video, and other elements. The electronic files that make up a website are stored on specialized computers called web servers. These computers accept requests from other (remote) computers for specific web pages and deliver those files needed to make the web page display on the remote computer. The type of program you would use to view web pages is called a web browser. It is this program that would make the requests to the web server for the website files.

EXAMPLE: There are websites on the World Wide Web dedicated to the films of famous actors.

Python: Python is a popular programming language created in the early 1990s by a man named Guido van Rossum.

Python is one of the most robust and versatile programming languages in existence. It is also considered by many to be one of the best programming languages for beginners to dive into.

CHAPTER TWO

The product of this chapter is a student that understands what coding is and the five elements of a program.

What is coding? Coding is computer programming. Computer programming is the act of creating computer programs, which are prepared sets of computer instructions designed to accomplish certain tasks. Programming consists of typing exact commands and instructions into a computer to create many of the things we use on computers on a day-to-day basis. To do computer programming, you must learn various programming languages and use these languages to create things on a computer that others can use.

EXAMPLE: The program on your computer entitled "calculator," which you can use to do math, is the result of computer programming.

Another word for these programs is "software." This is because the actual computer (the physical machine) is called "hardware." Therefore, the programs that run on that machine are called "software."

The terms "coding" and "development" mean the same thing as programming. A software developer is a computer programmer; so is a coder.

Programming is a spectacular thing because it is one of the few skills that apply to virtually all industries. Yes, companies that create software definitely utilize coding the most—but if you think about it, most industries utilize technology (software, websites, databases, etc.). And so, coders are valuable assets for companies in any industry—construction, food service, retail, transportation, etc.

There are many, many programming languages, and technology is ever-changing. The whole thing can be quite overwhelming, but there are basics that apply across the vast landscape of technology.

In general, the instructions in computer programs are executed in order, from the top down.

Five Elements

There are five key elements to any computer program:

1.   Entrance.
2.   Control/branching.

3.   Variables.
4.   Sub Program.
5.   Exit.

Let's take a look at each of these.

Entrance

A computer is a simple machine when you get down to it. It can only do one thing at a time, and it performs a computer program's instructions in the exact order in which the computer programmer puts them. It can only execute (perform or run) an instruction if it is directed to.

This means that any computer program has to have a clearly marked "first instruction." This is the first task that the computer will perform when the computer program is started. From that point forward, each instruction in the program will direct the computer what instruction to perform next after it performs the current instruction.

There are different ways to specify the entrance point, depending on which computer programming language is being used—but every computer program has a defined entrance point.

Control/Branching

Computers are often used to automate actions that would otherwise be performed by people. One of the most common things a person will be asked to do in performing a job is to assess the condition of a thing and, based on the condition of that thing, choose between two or more possible courses of action. In other words, they will have to make a decision. An example would be the activity of "a teacher grading a stack of papers":

• Take the next student paper from the top of the stack.
• Grade the paper.
• Write the grade on the paper.
• If the grade is 70% or higher, put the paper in a "Passed" stack.
• If the grade is below 70%, put the paper in a "Failed" stack.

You can see that there are two possible "paths" here. A path is "a possible course of action arising from a decision." Think of it as what happens when you come to a fork in the road. You have to decide on a course of action—which road do you take?

All but the simplest of computer programs will need to do the same thing. That is, they will have to check the condition of a piece of data, and based on the condition of that data, they will have to execute different sets of computer instructions.

In order to do this, the program will make use of special computer instructions called "control" instructions. These are just instructions that tell the computer what to look at in making a decision, and then tell the computer what to do for each possible decision. The most fundamental control statement for a computer is "if." It is used like this:

IF [condition to check checked] THEN [branch of computer instructions to execute]

Here, the "IF" is the control statement; the "THEN" is the branching instruction that points to the path of the program to execute if the control statement is true.

<u>Variables</u>

A variable is a piece of data that a computer program uses to keep track of values that can change as the program is executed. This might be something like "the grade of the student paper that was just graded" or "the color of paint to use for the next car on the assembly line."

Variables are a vital part of any computer program because they make it so a computer program can be used for more than a single, predetermined set of values. You can imagine that if "the color of paint to use for the next car on the assembly line" was only ever able to be "blue," the computer program using that data wouldn't be very useful. It would make a lot more sense to make it so the computer program could change that value for each car that was going to be painted.

When you are writing variables in a computer program, they usually are written in a manner like this:

[name of the variable] = [value of the variable]

For example, you might have something like this:

color = "red"

Here, the variable is named "color," and the value of that variable has been set to "red." In other words, the variable named "color" is now "equal" to the word "red."

Let's look at the example of "a teacher grading a stack of papers." Here, we could have a variable called "Paper Grade" that changed each time the teacher graded a paper. You could also have variables for the total number of questions on the paper ("Total Questions"), for the number of questions the student answered correctly ("Correct Questions"), and for the grade of the paper.

The written description from above:

- Take the next student paper from the top of the "To Be Graded" stack.
- Grade the paper.
- Write the grade on the paper.
- If the grade is 70% or higher, put the paper in a "Passed" stack.
- If the grade is below 70%, put the paper in a "Failed" stack.

In computer language, the procedure might look something like this:

1. Retrieve next Paper
2. Set Total Questions = [total questions in current Paper]
3. Grade paper
4. Set Correct Questions = [number of questions answered correctly]
5. Set Paper Grade = [Correct Questions/Total Questions]
6. If (Paper Grade >= 70%) then Paper Status = "passed"
7. If (Paper Grade < 70%) then Paper Status = "failed"

This is a simple computer program.

Each time the computer runs this program, it could have different values for each of the variables in the program, depending on how many questions the paper being graded has and how many of those questions the student answered correctly.

For example, let's say the paper has 100 questions, and the student answers 82 of them correctly. After the program is run, the result would be the following:
Total Questions: 100
Correct Questions: 82
Paper Grade: 82%
Paper Status: "passed"

Another example: Let's say the paper has 50 questions, and the student answers 30 of them correctly. After the program is run, the result would be the following:

Total Questions: 50
Correct Questions: 30

Paper Grade: 60%
Paper Status: "failed"

To clarify the need for variables: Let's say that at the time this computer program was being created, all papers at the school had 100 questions, and the teachers told the programmer to make it so that the number of questions was always assumed to be 100. In that case, the programmer wouldn't use a variable called "Total Questions." Instead, he could make the program look like this:

1. Retrieve next Paper
2. Grade paper
3. Set Correct Questions = [number of questions answered correctly]
4. Set Paper Grade = [Correct Questions/100]
5. If (Paper Grade >= 70%) then Paper Status = "passed"
6. If (Paper Grade < 70%) then Paper Status = "failed"

Notice that on line 4 of the program, the programmer set the number of questions to 100.

Now, let's say that the school introduces the concept of "quizzes," which are smaller papers with only 20 questions. Now, if the paper being handled by the computer program is a quiz, the grade will no longer be accurate—even if a student got all 20 questions correct, he/she would only get a grade of 20% (20/100).

A good programmer will analyze the need that the program is meant to resolve, then build the program so that it can handle changing aspects of that need over time.

Another valuable control statement is a loop. This is where part of the program is executed over and over until a certain condition is met.

In real-world terms, an example might be "grade papers one at a time until all the papers have been graded" or "make five copies of this document."

In a computer program, a loop would look something like this:

- [start loop]
  - o Perform action
  - o If [end condition has been met] then [exit the loop]
  - o If [end condition has not been met] then [repeat the loop]
- [end loop]

The program we looked at that grades papers could be set up as a loop. The instructions would be laid out like this:

- [start loop]
    o Take the next student paper from the top of the "To Be Graded" stack.
    o Grade the paper.
    o Write the grade on the paper.
    o If the grade is 70% or higher, put the paper in a "Passed" stack.
    o If the grade is below 70%, put the paper in a "Failed" stack.
    o If there are no more papers in the "To Be Graded" stack, exit the loop
    o If there are more papers in the "To Be Graded" stack, repeat the loop
- [end loop]

Often loops make use of a special variable called a "counter." The counter keeps track of how many times the loop has been executed. This can be used to make sure the loop is only executed when needed.

Let's add a counter to the grading program we're looking at as well as two new variables: "Total Papers" will be used to hold the value "how many papers need to be graded" and "Counter" will be used to hold the value "how many times has the loop been executed."

1. Set Total Papers = [total papers to be graded]
2. Set Counter = 0
3. If (Counter < Total Papers)
    a. Retrieve next Paper
b. Set Total Questions = [total questions in current Paper]
    c. Grade paper
d. Set Correct Questions = [number of questions answered correctly]
e. Set Paper Grade = [Correct Questions/Total Questions]
f. If (Paper Grade >= 70%) then Paper Status = "passed"
g. If (Paper Grade < 70%) then Paper Status = "failed"
h. Counter = Counter + 1
i. Go to step 3
4. [Continue on with the rest of the program]

Here, the loop is found in step 3.

Let's break down what each step is doing here:

Step 1: Count how many papers are in the "to be graded" stack and set the value of the "Total Papers" variable to that number.

Step 2: Create a variable called "Counter" and set it to the value zero. This variable will be used to keep track of how many papers are graded.

Step 3: Use the control statement "if" to see if we should execute a loop.

Step 3a–3g: Grade the paper; this has been covered above.

Step 3h: Since we have now graded a paper, add one to our Counter variable.

Step 3i: Go to the top of the loop, where we check to see if we need to execute the loop all over again.

Let's see what would happen if we used this program to grade two papers. Let's say that the papers look like this:

Paper 1:
Total questions on the paper: 100
Total questions that were answered correctly: 95
Paper 2:
Total questions on the paper: 20
Total questions that were answered correctly: 10

If we analyze what happens when the computer executes the program, it would look like this:

Total Papers = 2
Counter = 0
0 is less than 2, so loop will be executed
Paper 1 retrieved
Total Questions = 100
Paper 1 graded
Correct Questions = 95
Paper Grade = 95%
Paper Status = "passed"
Counter = 1
1 is less than 2, so loop will be executed
Paper 2 retrieved
Total Questions = 20
Paper 1 graded
Correct Questions = 10
Paper Grade = 50%
Paper Status = "failed"
Counter = 2
2 is not less than 2, so loop will not be executed
[Continue on with the rest of the program]

Sub Programs

As covered above, computer programs are generally executed in order, from the start point to the end point. This is called the "path of execution."

The main series of instructions in a program is called the "main program."

It is sometimes valuable to create another program that can be used by the main program as needed. This is called a subprogram. It is no different from any other program—it is made up of the same elements (entrance point, variables, control & branching statements, and exit point). However, a subprogram isn't used all by itself. Instead, the main program can execute the subprogram as needed. Here, the main program stops executing, and the subprogram starts executing. When the subprogram is done executing, the main program continues on where it left off.

This is called "calling" the subprogram—that is, the main program calls the subprogram, the subprogram starts and stops, and the main program continues on where it left off before calling the subprogram.

This is useful in creating programs because the computer programmer doesn't have to enter the instructions of the subprogram over and over; you only type them in once, and then when you need that subprogram to be called by the main program, you only have to type in one instruction—the instruction to call the subprogram. This lets you reuse the instructions you entered in for the subprogram rather than rewriting them.

Exit

Every program must have an instruction that tells the computer that the program is no longer running. Much like the Entrance, the exact instruction for this varies based on the computer language used, but all computer languages will have this type of instruction.

CHAPTER THREE

In this chapter, we will cover:

- Basic Python Terms,
- Computer Programming Basics, and
- Installing Python.

The product of this chapter is a student who is ready to begin writing code in Python.

There will be no coding in this chapter. We will be covering the terminology and basic coding data necessary to write code. While some of this may seem dry, knowing this information is vital for your success as a programmer.

Python is a programming language created in the late 1980s by Dutch computer programmer Guido van Rossum. Born in 1956, van Rossum worked for Google from 2005 to 2012, where he spent half of his time developing the Python language.

In the Python community, Van Rossum is known as the "Benevolent Dictator For Life" (BDFL). BDFL refers to the fact that he continues to oversee the Python development process, making decisions where necessary.

Due to the fact that Python resembles common English words and speech, it is considered by many to be one of the easier languages to learn. Python can be used to create apps, websites, games, and many other things. It is a very versatile language with many uses.

Python was named after the famous British TV show *Monty Python*, which aired from 1969–1974.

Python is used to build many popular web applications, e.g., Dropbox and BitTorrent. Python was even put to use in developing the search engines of YouTube, Yahoo, and Google.

Python-specific Definitions

There are some key definitions you must be familiar with to write code in Python. So, while it may be boring to read a whole bunch of definitions, to effectively write code in Python, you must understand the following words:

Iteration: To iterate means to say or do something again; to repeat something. Iteration is the act of repeating. Iteration means to go through a defined series of actions, repeating a certain number of times. Usually, this defined series of actions is repeated a certain number of times or until a condition is met.

EXAMPLE: Computer programs are usually created in iterations: Coming up with a basic working version, reviewing the program for mistakes to correct and improvements to make, doing that work, and repeating. This can be continued indefinitely.

Script: A script is a set of computer instructions that automates a task so that a multi-part task can occur without your involvement.

The origin of the term is similar to its meaning in "a movie script tells actors what to do" in that a script tells a computer what to do.

EXAMPLE: A script could be created that checks for new orders created at a manufacturing company every ten minutes and prints them off on a printer.

Syntax: Every spoken language has a general set of rules for how words and sentences should be structured. These rules are known as the syntax of that particular language. In the programming languages, syntax serves the same purpose. Syntax is the rules you must follow when talking to a computer and telling it what to do. There are many languages you can use to program a computer. Each language has its own syntax. Failing to use the syntax of a particular language correctly can mean that whatever you are designing will not work at all.

Syntax is the arrangement of words and phrases to create well-formed sentences in a language. In computer science, it is the language that allows the user to write out the program. Syntax is the rules about how to write code properly.

EXAMPLE: If a computer language required you write "cmd:" (meaning "command") at the beginning of each command, that would be part of the syntax of that language. And if you didn't write "cmd:" at the beginning of a command, the computer would not be able to process and execute the command because you violated syntax.

If you violated the syntax of Python, you could get an error message like this:

Traceback (most recent call last):

File "<pyshell#0>", line 1, in <module> print hello
NameError: invalid syntax

This could occur for things like missing a colon after a conditional statement, improper indentation, using the wrong operator, etc.

Parse: To break something up into its parts and analyze it. In computing, it means that the program code is analyzed and read.

EXAMPLE: If you write code for a website, the web server (computer that gives and receives data over the Internet) parses the code and then outputs the correct HTML (popular programming language for websites) code to the browser (program one uses to display the Internet).

Parsing can also refer to breaking up ordinary text. For example: Search engines (programs used to search the web) typically parse search phrases entered by users so that they can more accurately search for each word.

Parser: A parser breaks code into parts and analyzes it. It's a tool that examines program instructions while still in high-level language and analyzes their syntactic structure. Parsers receive instructions and break them into parts, routing each part to the correct part of the compiler/interpreter for handling. Parsers also look over information received to ensure it is complete and no vital components are missing.

EXAMPLE: If a parser finds a syntactic error, it will report back an error message.

IDLE: Integrated Development Environment (IDE). An IDE is a set of programming tools for writing software programs.

Where most IDEs can be used to write programs in several different computer languages, IDLE is only used to write programs in the popular computer language Python.

Guido van Rossum (the creator of Python) says IDLE stands for "Integrated Development Environment"—but since van Rossum named Python in honor of the British comedy group Monty Python, the name IDLE was probably chosen partly to honor Eric Idle, one of Monty Python's founding members.

IDLE is intended to be a simple development environment suitable for beginners, especially in an educational setting.

EXAMPLE: You can get IDLE for free if you want to start creating computer programs using the programming language Python.

Shell: The term "shell" in computing—being the outer layer between the user and the operating system—means the same thing(s) as shell does in normal English. It is a program that gives an interface for the user to use to issue commands to the operating system.

EXAMPLE: Code is often written in a shell.

In Python, the Shell and IDLE look like this:

Python also includes a text editor that looks like this:

Tar: Tar is a type of file format as well as the name of a program used to handle such files. The name comes from the two words "tape archive."

Tar is now commonly used to collect many files into one larger file for distribution or archiving while preserving file system information such as user and group permissions, dates, and directory structures.

In order to reduce storage requirements, tar files can be compressed. The most common program used to compress tar files is a compression program called gzip. Tar files that have been compressed by the gzip program have the file extension "tar.gz." In order to access the tar file, the gzip program must be used to decompress the tar file. There are other decompression programs available that are capable of decompressing .gz files.

Operators: In mathematics, an operator is a symbol used to carry out a computation. There are several different kinds of operators.

Arithmetic operators such as +, -, /, *, % are used to perform math functions. Operators such as >=, ==, != are used to compare values. Logical operators like "and," "or," and "not" are used to evaluate whether an expression is true or false.

EXAMPLE: In creating programs, instructions like "AND" and "OR" are considered operators. For example, an instruction might specify:

If someone types in both "John" AND "Sally," then turn the screen blue

(Here, AND is an operator. It is a symbol for the action of comparing whether two things are both true.)

Another instruction might specify:

Total price = base price + tax

(Here, + is an operator. It is a symbol for addition).

Other common operators are the mathematical actions for multiplication or division. The symbols used for these operations are usually an asterisk (*) for multiplication and a forward slash (/) for division.

EXAMPLES:

AgeInMonths = Years * 12 (* is an operator).

PricePerItem = TotalPrice / NumberOfItems (/ is an operator)

Array: A collection of data arranged in rows and columns. In computers, an array is a group of related things that are stored together in a sequence. It is a logical way things can be organized in a computer. Arrays can be quite simple or quite complex.

EXAMPLE: A simple array would be something like the numbers 7, 3, and 15. It would be written out like this:

[7, 3, 15]

These three pieces of data are called elements—they are the elements of the array.

A system is needed for identifying each element of an array. The simplest method for this is to start numbering them at zero, starting at the left position, and count up from there.

In the above example, the element "7" would be at position 0; "3" would be at position 1; and "15" would be at position 2.

Another word for the position of an element is the "index" of the element—for the above example of an array, index 0 is "7"; index 1 is "3"; etc.

Each element, therefore, has two properties: its index and its value.

EXAMPLE: You have three pictures of your cat, and you could save them in an array: CatPic1, CatPic 2, and CatPic 3. Here, index 1 has a value of "CatPic2."

Integer: In order to understand the term "integer," you will first need to understand the terms "memory," "variable," "declare," "assign," and "data type."

Memory is a physical device used to store information on a computer. It is recorded electronic data that can be accessed in the future. A memory device could be thought of as a collection of boxes called "memory locations." A memory location has two aspects: a unique identifier, called an "address," and the actual data stored in that location.

A variable is a construct (an idea formed from simpler elements) used to store data that may change as the computer performs its tasks. It is three elements working together: a memory location, an associated symbolic "identifier," and a "value"—the data actually stored in the memory location. In other words, "this is where we are storing the data (*memory location*), this is how to refer to the location (*identifier*), and this is the data stored in that location (*value*)."

EXAMPLE: A variable could have an identifier of "age," have a value of "27" and be stored at memory location 1000.

Note that there is another common term for the identifier: the "name" of the variable. In the above example, the name of the variable is "age."

When you create computer programs, you will often make use of variables. This involves two actions:

First, designating a particular memory location and an associated identifier. This is called "declaring" the variable.

Second, setting the value of the data to be stored in the memory locations. This is called "assigning" the value to the variable.

A data type is a distinct type of data used by a computer. There are several data types. A data type includes the type of information represented by the data as well as the behavior of the data. This information tells the computer what kind of information it is working with and what things it can do with that information.

An integer is a data type that represents whole numbers (numbers that don't have any fractions). This could be numbers like 5, -8, 1024, etc. The number 4.5 would not be an integer because it includes a fraction of a number in its value. The behavior of this data type includes mathematical operations like addition, subtraction, etc.

An integer variable would be declared and assigned like this:

integer [variable name] = [whole number]

EXAMPLE: You could have a computer keep track of the number of students in a class using a variable called "classSize." You could set it to 25 like this:

integer classSize = 25

String: A string is a data type that represents text—that is, letters, numbers, and other symbols. The behavior of this data type includes things you would do with text—add on more symbols, convert the text to uppercase, etc.

A string variable would be declared and assigned like this:

string [variable name] = "[set of alphanumeric characters]"

EXAMPLE: You could have a computer keep track of the name of a student using a string variable called "studentName." You could set it to "Jane Smith" like this:

string studentName = "Jane Smith"

Concatenate: To connect things together, like links in a chain.

It means to take one piece of data and stick it on the end of another piece of data.

EXAMPLE: Concatenating the text "device" and the text "s" makes the text "devices."

Expression: Numbers and symbols and operators grouped together that show the amount of something. An expression is a written math problem.

In computers, an expression is a combination of values that are computed by the computer. There are different ways to write out expressions, depending on which language you are programming in.

EXAMPLE: Name = "What person types in the name box" could be an expression. Also: "x + 5" is an expression.

Function: A function is another word for a subprogram.

There are other words in technology for this concept: method, subroutine, subprogram, procedure, routine.

Statement: Statements are computer instructions. These are the instructions that are used by people as they create computer programs. The simplest of these might be things like "print," "delete," "add," "multiply," etc.

EXAMPLE: The "print" statement tells your computer to print whatever text you typed as part of the command.

Call: To "call" means to demand or direct something. In normal English, this could be used like, "This calls for celebration!" In computers, a call is a direction by a main computer program to execute the tasks of a subprogram. More specifically, a "call" is when a program temporarily transfers control of the computer to a subprogram. Once the subprogram is done executing, control of the computer is returned to the main program. A program could make many "calls" to multiple subprograms as the program does its sequence of tasks.

EXAMPLE: If you were using a video game program, the video game program could call a "high score" subprogram after every game ended to make the words "High Score" pop up on the screen.

Parameter: Sometimes the subprogram needs some information from the main program in order to perform its tasks. When the subprogram is created, its description might include this information. That information is called the "parameters" of the subprogram.

Often, the subprogram will do its work and then provide the main program with some information derived from its work. That is called returning information to the main program.

EXAMPLE: You might have a subprogram that adds two numbers. It could look like this:

```
subprogram AddTwo(x, y)
{ return x + y }
```

Here, the name of the subprogram is "AddTwo." The parameters of the subprogram are two numbers called x and y. The subprogram will return the sum of those two numbers.

The creation of a subprogram, as shown above, is called "defining" the subprogram. The above is an example of a subprogram definition.

Let's look at how that would be used in a main program. Say you had a school with two Physical Education classes per day, and you wanted to have the program calculate the total number of students in those classes. You could make a main program that made use of our "AddTwo" subprogram. It might look like this:

```
classSizePE1 = 25
classSizePE2 = 43
totalSize = AddTwo(classSizePE1, classSizePE2)
print "The total number of students in the PE classes is: " totalSize
```

Let's look at this program one line at a time.

Line 1: The main program created a variable called "classSizePE1" and gave it a value of 25.

Line 2: The main program created a variable called "classSizePE2" and gave it a value of 43.

Line 3: There are five things happening here.

First, the main program created a variable called "totalSize."

Second, the main program called the subprogram "AddTwo." That subprogram was given the values 25 and 42 since those are the values of the variables "classSizePE1" and "classSizePE2."

Third, the subprogram "AddTwo" performed its work, taking the values 25 and 43 and adding them together to create the value 68.

Fourth, the subprogram returned that value to the main program.

Fifth, the main program gave that value to the variable "totalSize."

Line 4: The main program then continued running. Here, the program used the value of the variable totalSize to display this text on the screen:

The total number of students in the PE classes is: 68

This brings us to the concept of arguments. Arguments are the actual data passed to a subprogram when it is called. In the above example, the arguments are the numbers 25 and 43.

This comes from mathematics. In mathematics, there are formulas. These are exact math operations that are done in an exact order. Typically, these math formulas need to be given some initial values to start processing the math steps. Those initial values are called arguments.

A subprogram doesn't necessarily need any arguments. Some subprograms may take one argument; some may take more than one.

To clarify: when a subprogram is defined, any data items it will need are called parameters. When the subprogram is actually used, the actual data passed to it at that time is called arguments.

<u>Installation</u>

Now you can install Python!

1.   Go to this link:

https://www.python.org/downloads/

2.   Find the most recent version of Python 2.

3.   Download and install the newest version of Python 2.

Note: There are two primary versions of Python – 2 and 3. They're very similar to one another and at the time this book was written, Python 2 was more popular.

If you run into any difficulty downloading and installing Python on your PC or Mac, research online for solutions, tutorials, etc.

Once Python is downloaded and installed, we can start writing code!

CHAPTER FOUR

In this chapter we will cover:

- Writing basic Python code
- Variables
- Data types
- Scripts
- Commenting
- Notation
- Strings
- Dates
- String functions

The product of this chapter is a student who has successfully written and executed basic Python code.

Open up IDLE. There are several ways to do so, depending on your machine. One way is to search your computer for IDLE and then run it.

Once IDLE is open, type the following in IDLE:

```
>>>print 'Hello world'
```

and press enter.

Notice that "Hello world" is now printed. This is called *returning a value*. The value that has been returned, in this case, is the string "Hello world."

NOTE:

1. >>> indicates where to type commands in IDLE.

Type the following in IDLE:

```
>>>print Hello world
```

(without using quotation marks) and press enter.

Notice that there is an error. Quotation marks tell the computer that you are entering a string. Without the quotation marks, the computer is looking for a variable or built-in command that has not been defined.

Variables

Type the following in IDLE:

```
>>>J = 50
```

and press enter. You have just created a variable. You assigned a value of 50 to the variable J.

Now type the following in IDLE:

```
>>>print J
```

and press enter.

Notice that the value 50 is returned. In this case, you didn't need to include the quotation marks because you were printing a variable that has been defined.

Type the following in IDLE:

```
>>>print j
```

(lowercase j)

Notice that we get an error. This is because Python is case-sensitive. You assigned your variable as J (capital).

Now type the following in IDLE:

```
>>>Print "Hello there!"
```

(capital P) and press enter. We again get an error. This is because the print command is supposed to be lower-case.

Math

We can add numbers in Python. Type the following in IDLE:

```
>>>1 + 1
```

and press enter. The value returned is 2.

Now type the following in IDLE:

```
>>>K = 5
>>>print (J+K)
```

and press enter. Notice the value returned is 55. That is because J (50) + K (5) = 55.

As a note, you could have written K=5 (without spaces) in the above code, and it would still run.

We can subtract numbers in Python too. Type the following in IDLE:

```
>>>2-1
```

and press enter. The value returned is 1.

```
>>>print (J-K)
```

and press enter. Notice that the value returned is 45.

We can even multiply numbers in Python. Type the following in IDLE:

```
>>>4*5
```

and press enter. The value returned is 20.

Now type the following in IDLE:

```
>>>print (J*K)
```

and press enter. Notice that the value returned is 250.

Lastly, we can divide numbers in Python. Type the following in IDLE:

```
>>>100/5
```

and press enter. The value returned is 20. Now type the following in IDLE:

```
>>>print (J/K)
```

and press enter. Notice the value returned is 10.

Data Types

The most common types of data in Python are:

1) String   (A sequence of characters surrounded by quotation marks)
2) Float   (A number with a decimal point)
3) Boolean   (A value that can be either true or false)
4) Integer (Int – a whole number)

Create a string variable by typing the following in IDLE:

```
>>>A = "John is happy"
>>>A
```

and press enter. As a note, for the most part, a quotation mark (") and an apostrophe (') are interchangeable in this version of Python.

"A," the variable, returns the string.

To create a float variable, type the following in IDLE:

```
>>>B = 10.75
>>>B
```

and press enter. "B," the variable, returns the float.

Boolean

A Boolean is a data type that has only two possible values: "true" and "false." It is used in computer operations that compare one or more conditions. This is a key tool in using computers to perform different sets of operations based on the condition of certain pieces of data.

It is based on Boolean logic, a form of logical analysis in which the only possible results of a decision are true and false. This logical system is based on the work of George Boole, an English mathematician, educator and philosopher.

A Boolean variable would be declared and assigned like this:

Boolean [variable name] = [*true* or *false*]

EXAMPLE: You could have a computer keep track of whether or not a specific employee is allowed to access the company bank accounts with a

Boolean variable called "AllowedFinancialAccess." You could set it to "false" like this:

Boolean AllowedFinancialAccess = false

Booleans are a vital part of computer programming since they are used with computer instructions that compare conditions and then execute one of two or more possible actions based on the comparison.

To utilize boolean logic, type the following in IDLE:

>>>J is 50

and press enter. As you can see, *True* is returned. That is because earlier, J was assigned the value 50.

Now type the following in IDLE:

>>>J is 25

and press enter. As you can see, *False* is returned.

Type the following in IDLE:

>>>J is not 50

and press enter. As you can see, *False* is again returned.

The symbol ">" means "greater than." We can utilize this in Python. Type the following in IDLE:

>>>25 > 10

and press enter. We are stating that 25 is more than 10, which is true.

Now type the following in IDLE:

>>>10 > 25

and press enter. We are stating that 10 is more than 25. This is false.

The symbol "<" means "less than."

Type the following in IDLE:

```
>>>25 < 10
```

and press enter. We are stating that 25 is less than 10, which is false.

Now type the following in IDLE:

```
>>>10 < 25
```

and press enter. We are stating that 10 is less than 25, which is true.

The symbol "==" means "equal to." We use two equal signs to designate that we are checking for equality as opposed to assigning a value with a single equal sign (as in x=5).

Type the following in IDLE:

```
>>>10 == 10
```

and press enter. We are stating that 10 is equal to 10, which is true.

Now type the following in IDLE:

```
>>>25 == 10
```

and press enter. We are stating that 25 is equal to 10, which is of course false. These last few lines of code we've written are all boolean logic.

Commenting Code

Commenting code is the act of leaving notes in your code that explain various things for yourself and/or others. The comments are not run in the program but are read by people when viewing the code directly so as to understand the code better.

To write comments in Python, you simply add "#" at the beginning of the statement.

In IDLE, write the following:

```
>>>25 > 20 #Here I am stating that 25 is greater than 20, which is true
```

and press enter. The program ran, and your comment wasn't displayed—but now someone in the future knows the purpose behind your code. It is good practice to comment one's code, as it helps others understand your logic.

Scripts

Now let's write a script. In IDLE, go to File and open a New File. This allows you to be working in the text editor creating a program to run rather than working directly in the shell.

Type the following in the text editor:

```
J = 50
K = 5
print (J+K)
```

Now we will run the program. First, you must save the file,save your Python files with the ".py" extension; this tells the computer that the file contains Python code). Once it's saved, click Run, and then Run Module (F5 on Windows and fn+F5 on Mac).

The program should run and return the value 55.

If Statements

Type the following in the text editor:

```
if J > 25:
    print 'J is the winner!'
```

Run the program. The program will now print "J is the winner!" because the "if" statement (a boolean statement) evaluated to true because J is greater than 25. Note: If you deleted your earlier code where you assigned J as 50 (J = 50), the code won't run.

Indentation

In Python, you have to indent your code when a statement has been made. As you may have noticed when writing the above *if* statement, this is typically indented automatically for you.

Let's try this out. Type the following in the text editor:

```
if B == 20:
    print '20 is equal to B'
```

Did you notice how the code was automatically indented? (Note: indents are also caused by pressing Tab.)

Now, run the code. You have gotten an error message. This is because you need to define the variable B.

Try assigning the variable "B" the value "20" by writing the following in the text editor above your code:

```
B = 20
```

Run the code again, and it works!

Strings

In the text editor, click Run and then Python Shell (open up a new Python Shell Window). Type the following in IDLE to create a variable called "name" (NOTE: You can type your own name inside the quotation marks. The following is just an example):

```
>>>name = 'Ashley'
>>>print name.lower()
```

and press enter. Your name is now displayed in lower-case. Note: If you are simply copying and pasting your code into IDLE instead of typing it, you will occasionally get errors. For example: the above single quotation marks (' ') will not run in IDLE because the Python single quotation marks (' ') are straight up and down. As we mentioned earlier, ensure you write all code yourself and don't copy and paste.

Now type the following in IDLE:

```
>>>print name.upper()
```

and press enter. Your name is now displayed in capital letters. We used a built-in function in Python.

Type the following in IDLE:

```
>>>print name[0]
```

and press enter. Then type the following in IDLE:

```
>>>print name[2]
```

and press enter again.

Can you see what this code does? You are printing characters from a string that exist at a certain index (see the definition of "array" if you need a reminder of what an index is). Indexes typically start at 0 in programming.

Lists

Let's create a "list." Type the following in IDLE:

```
>>>listA = ['mother', 'father', '1970', '1965'];
>>>print listA[3]
```

and press enter. In this code, you have created a list and then asked for the fourth item in the list to be displayed.

Tuples

Another type of list is called a "tuple." The difference between a "list" and a "tuple" is that tuples can't be altered, and tuples use parentheses instead of square brackets. Tuples are immutable data types. "Immutable" means "unable to be changed."

Let's create a tuple. Type the following in IDLE:

```
>>>tupA = ("brother", "sister", "1990", "1992", "1993", "1995")
>>>print tupA[1:4]
```

and press enter. If you notice, that prints the items, starting at index 1 to (but not including) index 4.

Dates

Let's set a date. Type the following in IDLE:

```
>>>date = "3/13/2020"
```

and press enter.

We are going to use the .split() command to split apart the day, month, and year.

Enter this code in IDLE:

```
>>>split_the_date = date.split('/')
>>>print split_the_date
```

and press enter (you have to include a print statement or the date won't display). We have used the date.split('/') command to separate the parts of the date.

Enter the following in IDLE:

```
>>>print split_the_date[0]
>>>print split_the_date[1]
>>>print split_the_date[2]
```

and press enter.

Concatenating a String

We will now concatenate (see definition of "concatenate" if needed) strings so that the dates we entered are displayed with the day, month, and year.

Enter the following code in IDLE:

```
>>>print "Month: " + split_the_date[0]
>>>print "Day: " + split_the_date[1]
>>>print "Year: " + split_the_date[2]
```

and press enter.

Now, open a New File. Enter the following code in the text editor:

```
date = "12/25/2025"
full_date = date.split('/')
print ('Month: ' + full_date[0] + ' Day: ' + full_date[1] + ' Year: ' + full_date[2])
```

Run the code.

Functions

You can change the cases of letters through using the .swapcase() function. Click on Run and then Python Shell, and we are back in IDLE. Type the following in IDLE:

```
Name = "Erik"
Name.swapcase()
```

Run the code.

You can also get rid of spaces, indents, and tabs from a string by using the .strip() function.

Enter the following code:

```
My_Favorite_Color = "   My favorite          color   is   blue        "
print My_Favorite_Color
My_Favorite_Color.strip()
```

Note: In our code, we put a tab between each word instead of a space—this can be done by holding Ctrl and Tab.

Run the code. You should see the empty space at the ends and beginnings of the string eliminated.

<u>Using What We've Learned Thus Far</u>

Open up the text editor and write the following code:

```
B = "Let's do some math!"
print B
X = 10
print "X = 10"
print "X plus 5 equals:"
print X + 5
A = 5
print "A = 5"
print "A minus X equals:"
print A - X
print "X times A equals:"
print X * A
print "X divided by A equals:"
print X / A
print "Is X larger than A?:"
print X > A
print "Is X less than A?:"
print X < A
print "Are X and A equal?:"
print X == A
if X > A:
    print "X (being 10) is larger than A (which is 5)"
name = 'BILLY'
print name
print "Now it is time to print BILLY in all lower case letters: " + name.lower()
list_names = ['Billy', 'Sally', 'Johnny', 'Raphael']
```

```
print "Here's the list we created: "
print list_names
print "Here's the third name from the list we wrote in all caps: " +
list_names[2].upper()
date = "July/14th/1987"
print "Here's the date we created: " + date
split_date = date.split('/')
print "Here's the date we entered split apart: "
print split_date
another_name = 'DiAnE'
print "We chose the name: " + another_name
print "Here's " + another_name + " written with the cases swapped: " +
another_name.swapcase()
```

Run the code. In this code, we have assigned variables, used functions, implemented *if statements* and other skills you've learned up to this point.

Build Your Own

Here's a challenge for you! Make a program that does the following:

1. Prints something
2. Assigns a variable
3. Performs a math function
4. Uses an *if statement*

End of Chapter Quiz

Can you answer the following questions?
1. What is a variable?
2. What are some data types in Python?
3. What is a script?
4. Why is it important to comment code?

CHAPTER FIVE

In this chapter, we will cover:

- Statements
- Loops
- Lists
- Dictionaries
- Functions

The product of this chapter is a student who is familiar with some of the most-used Python code.

Not Equal To (!=)

In Python, the operator "!=" means "check that the value on the left is not equal to the value on the right." Let's use this. Open up IDLE and type the following:

```
>>>10 != 10
```

and press enter. The statement that 10 is not equal to 10 is False.

Now type the following in IDLE:

```
>>>10 != 9
```

and press enter. The statement "10 is not equal to 9" is true.

Greater Than or Equal to (>=)

The operator ">=" means "check that the value on the left is equal to or greater than the value on the right." Let's try this out.

Type the following in IDLE:

```
>>>10 >= 10
```

and press enter. The statement "10 is greater than or equal to 10" is true.

Now type this in IDLE:

```
>>>11 >= 10
```

and press enter. The statement "11 is greater than or equal to 10" is true.

And now type the following in IDLE:

```
>>>5 >= 10
```

and press enter. The statement "5 is greater than or equal to 10" is false.

Less Than or Equal to (<=)

The operator <= means "check that the value on the left is less than or equal to the value on the right." Let's try this out. Type the following in IDLE:

```
>>>10 <= 10
```

and press enter. The statement "10 is less than or equal to 10" is true. Now type this in IDLE:

```
>>>5 <= 10
```

and press enter. The statement "5 is less than or equal to 10" is true. And now type the following in IDLE:

```
>>>11 <= 10
```

and press enter. The statement "11 is less than or equal to 10" is false.

Testing Variables

Type the following in IDLE:

```
>>>A = 10
>>>A > 7
```

and press enter. This is true: A (which has been assigned the value of 10) is greater than 7.

Repeat this exercise with all of the following symbols in IDLE:

```
<       (less than)
>=      (greater than or equal to)
<=      (less than or equal to)
```

```
==      (equal to)
!=      (not equal to)
```

Another If Statement

We covered *if statements* earlier. Let's do some more. Type the following into IDLE:

```
>>>A = 10
>>>if A != 5:
        print "A is not equal to 5."
```

Else Statements

Else statements are statements used to cover a potential outcome other than what was laid out by one or more "If" statements. "Else" statements contain a block of code that executes if the conditions in the "if" statements are not met.

Let's use an Else Statement.

Open up a New File (Click File > New File). Type the following in the text editor and then run the program:

```
name = "Jack"
if name == "Jack":
    print "Hello, Jack"
else:
    print "You are not Jack"
```

Now change the name (and then re-run the program) such as this:

```
name = "Emily"
if name == "Jack":
    print "Hello, Jack"
else:
    print "You are not Jack"
```

That is how Else Statements work. Again, notice the indentation of the code.

Elif Statements

"Elif" statements literally mean "Else if." They are additional "If" statements that cover additional options. Let's write one. Delete the code in the text editor and type the following:

```
answer = "Red"
if answer == "Red":
    print "You chose red."
elif answer == "Blue":
    print "You chose blue."
else:
    print "Please enter Red or Blue as your answer."
```

Run the code.

Now change *answer* to "Blue" and run the code again.

And now change *answer* to "Yellow" and run the code again.

Exercise

Write your own program that contains an If statement, Elif statement, and Else statement. You can keep it simple at this point.

Counter

A counter keeps track of how many times something occurs. As the name implies, it counts.

Open the text editor and enter in the following code:

```
counter = 0
print counter
counter = counter +1
print counter
counter = counter +2
print counter
counter = counter +3
print counter
counter = counter -6
print counter
```

Now run the code. The program counted and displayed 0, 1, 3, 6, and 0. Now, let's slow it down so the numbers don't all display at once ...

Time.Sleep()

Time.Sleep() allows you to delay the running of certain blocks of code by specified seconds or portions of seconds (as in a float).

In order to use this properly, we'll have to cover an important concept in Python: Modules.

A module is a pre-made collection of Python code for your use. Python has many modules that exist to assist developers. These modules have functions you can use to add functionality to your program.

Some functions come built into the Python language (like .upper() and .lower(), etc.), but other functions require you to import a module before they can be used. Importing a module just means adding the code of that module to your Python program so that the program can use it.

To use the Time.Sleep() functions, you must first import the "time" module.

Enter this at the top of the code you just wrote (before everything else):

```
import time
```

Now enter the following after every print statement in your code:

```
time.sleep(.5)
```

Your code should look like the following, and you should see a pause between each counter display:

```
import time
counter = 0
print counter
time.sleep(.5)
(etc.)
```

Run your code. Your code should now display with slight delays between each printed counter.

Loops

The "range() function" generates a list of numbers. The "For Loop" repeatedly executes instructions as long as a particular condition is true. Write the following code in the text editor:

```
import time
for counter in range(1,11):
    print counter
    time.sleep(.5)
```

Run this code. Or, to count down, enter and run the following code:

```
import time
for counter in range(10, 0, -1):
    print counter
    time.sleep(.5)
```

Do you see how this code works? The range function takes three parameters (start, stop, and step). In the above, we are stepping through the range by -1. Loops make your code more efficient and can save you time.

The While Loop means that "while (blank) is occurring, do (blank)."

Open a New File inside IDLE and enter in the following code:

```
import time
counter = 1
while counter < 11:
    print counter
    time.sleep(.5)
    counter = counter + 1
    time.sleep(.5)
```

and run this code.

Now delete the "counter = counter + 1" and run the code. As you can see, without exact instructions, the loop will continue on forever, and we do not want to get stuck in an infinite loop. You can stop it by exiting the program.

Lists

Lists are written like this: Name_of_list = [item1, item2], etc.

To make a list, type the following into a New File in the text editor:

```
Best_bands_list = ["The Beatles", "Rolling Stones", "Led Zeppelin", "Jimmy Hendrix"]
print Best_bands_list
```

and run the program (as a note, it is common for programmers to use _ instead of a space between words).

Now, to print text with the list, try this:

```
Best_bands_list = ["The Beatles", "Rolling Stones", "Johnny Cash", "Beach Boys"]
print Best_bands_list + "These are the best bands!"
```

and run your code.

As you can see, you got an error that states: "can only concatenate list (not 'str') to list."

We are having trouble concatenating the two different data types (a string and a list). To repair this error, we can specify which string item from the list we want to concatenate with the string by typing and running the following:

```
Best_bands_list = ["The Beatles", "Rolling Stones", "Johnny Cash", "Beach Boys"]
print Best_bands_list[0] + ' is an awesome band!'
```

The same can be done with the remaining items of the list by calling out each by their index number: [1] for Rolling Stones, [2] for Johnny Cash, and [3] for Beach Boys.

Changing Lists

To change a list, we specify which item we want to replace or change. First, let's add something to the list. Type the following at the end of your code:

```
Best_bands_list.append("The Temptations")
print Best_bands_list
```

and run the code. We appended a new band onto the list (The Temptations), which is index number [4].

Now let's replace an item on the list. Type the following at the end of your code:

```
Best_bands_list[2] = "James Brown"
print Best_bands_list
```

and run the code. We removed Johnny Cash and added in James Brown in his place.

Lists and Loops

We will now combine a list with a "For Loop." Open a New Window and type the following into the text editor:

```
import time
Best_bands_list = ["The Beatles", "Rolling Stones", "Led Zeppelin", "Jimi Hendrix"]
for band in Best_bands_list:
    print band + " are a great band!"
    time.sleep(.5)
```

and run the code (note: "band" is the variable name for the above indexes when iterating through the list. You could name the variable anything: x, y, number, etc).

Numbers and Lists

Let's create a list with numbers. We will be multiplying these numbers to the power of ten (note: ** is what you type to find out the power of a value).

Open a New File and type the following into the text editor:

```
List_of_numbers = [5,10,15,20,25]
for numbers in List_of_numbers:
    print numbers**10
```

and run this code.

Now, let's figure out how to store a value from our previous code into a new list. Create an empty list by typing the following code into the text editor:

```
import time
Values_list = []
```

```
Values_list.append(numbers**10)
print Values_list
```

and run the code.

We just created and displayed a list that contains 25 to the power of 10.

Dictionaries

In computer programming, dictionaries are a specialized type of list. The first item is the key, and the second item is the value. Together, they are a "key-value pair."

Open up IDLE and type the following:

```
>>>Birthday_Dictionary = {'Emily' : 'June 1950' , 'Maxine' : 'March 1962' , 'Kelly' : 'May 1955' , 'Violet' : 'January 1948'}
```

and press enter. You have created a dictionary. The most common way to create a dictionary is the way we did above (utilizing curly brackets {}). Now let's view the dictionary. Type this:

```
>>>print Birthday_Dictionary
```

and press enter.

Key-value pair is abbreviated "KVP." These are sets of two linked data items that consists of: a key, which is a unique identifier for some item of data, and the value, which is the data that is identified by the key. The key is the unique name, and the value is the content.

Collections of Key-Value Pairs are often used in computer programs.

EXAMPLE: Here is an example of a collection of KVPs that might be used in a computer program for a school. Here, the KEY is used to store the name of a course, and the VALUE is the description of the course:

KEY	VALUE
ALG 1	"Algebra 1"
ALG 2	"Algebra 2"

HIS	"History"
PHYS1	"Physics 1"
PE	"Physical Education"

Note that in the above list, you could not have a second Key-Value Pair that used the Key "HIS," as the keys in a given collection of KVPs must be unique.

In the above dictionary, the name is the key, and the birth date is the value. Now let's find someone's birthday by tying this:

```
>>>print Birthday_Dictionary['Maxine']
```

and press enter. We are given the value "March 1962" —Maxine's Birthday.

To replace someone's birthday, type the following in IDLE:

```
>>>Birthday_Dictionary['Emily']='December 1975'
>>>print Birthday_Dictionary
```

pressing enter for each line.

To add text in combination with displaying the Dictionary, type the following in IDLE:

```
>>>print "Kelly's Birthday is: " + Birthday_Dictionary['Kelly']
```

and press enter.

To add another birthday to the dictionary, type the following in IDLE:

```
>>>Birthday_Dictionary['Sara'] = 'November 1980'
```

and press enter.

Then type this in IDLE:

```
>>>print Birthday_Dictionary
```

and press enter. You can see the earlier changed birthday and the newly added person and birthday.

To delete something from a dictionary, type the following into IDLE:

```
>>>del Birthday_Dictionary['Kelly']
```

and press enter. Then display the dictionary to verify Kelly's entry was taken out:

```
>>>print Birthday_Dictionary
```

and press enter.

Functions

As we covered earlier in this book, a function is a block of organized, reusable code that is used to perform a single, related action. A call is a direction by a main program to execute the tasks of a function (a set of instructions written inside a program that perform some tasks that you may want to do over and over again at various times).

To create a function, open a New File and type the following in the text editor:

```
def Subtraction(A,B):
    subtract = A - B
    return subtract
print Subtraction(20,10)
```

and run the program. "def" means we are defining your function. A and B are the variables.

Now add the following code:

```
print Subtraction(10,5)
print Subtraction(500,350)
```

and run the program.

Look over what we just did with your code. We just created a function with two parameters that will automatically subtract the two values that we enter in as arguments. This function automatically handles any subtraction we want to do.

The best way to understand functions is to create them. Let's create another function. Type the following in the text editor:

```
def addition(C,D,E):
    add = C + D + E
    return add
print addition(5,10,20)
```

and run the code.

Now we will create a multiplication function. Type and execute the following in the text editor:

```
def multiplication(F,G,H):
    multiply = F*G*H
    return multiply
print multiplication(2,4,6)
```

Now we will create a division function. Type and execute the following in the text editor:

```
def division(I,J,K):
    divide = I/J/K
    return divide
print division(20,4,2)
```

Did you notice how your division returned a value of 2 when technically it should have been 2.5? That is because we are in integer mode. To allow floats (decimal) in what we are doing here, we tweak our code in the text editor to the following:

```
def division(I,J,K):
    divide = float(I)/float(J)/float(K)
    return divide
print division(20,4,2)
```

and run the code.

String Function

There are many functions already built into Python. One of these functions is the string function. This function is used to convert any data type to a string. Open IDLE and type the following code in:

```
>>>X = 500
```

```
>>>print str(X)
```

pressing enter after each line. Notice that the function has taken the integer type 500 and converted it to text. That text is then displayed on the screen.

Float Function

The float function ensures that integers divided into decimal numbers are properly displayed. Type the following in IDLE:

```
>>>K = float(699)/float(222)
>>>print K and press enter.
```

pressing enter after each line.

Length Function

The length function gets the length of whatever you are typing in. Enter the following in IDLE:

```
>>>How_many_characters = len("I am learning a lot from this Python book!")
>>>print How_many_characters
```

pressing enter after each line. This returns back the length (how many characters) of the string "I am learning a lot from this Python book!"

Integer Function

To change a float into an integer, use the integer function. Type the following in IDLE:

```
>>>Z = float(219)/float(67)
>>>print Z
>>>Whole_number_Z = int(Z)
>>>print Whole_number_Z
```

pressing enter after each line.

Round Function

To round a number to the nearest whole number, use the round function. Type the following in IDLE:

```
>>>X = 29.76521
>>>print round(X)
```

pressing enter after each line.

Delete

To delete items, type the following in IDLE:

```
>>>X = 15
>>>print X
>>>del X
>>>print X
```

pressing enter after each line. As you can see, you get an Error message after you delete because X is no longer stored.

Using What We've Learned Thus Far

Open up the text editor and write the following code:

```
X = 10
print "X (10) is not equal to 15:"
time.sleep(.5)
print X != 15
print "X is greater than or equal to 15:"
time.sleep(.5)
print X >= 15
if X <= 15:
    print "X is not equal to or greater than 15."
    time.sleep(.5)
number = 25
if number == 15:
    print "25 is not equal to 15."
else:
    print "The number is 25."
    time.sleep(.25)
counter = 25
print counter
counter = counter + 5
time.sleep(.25)
print counter
time.sleep(.25)
counter = counter - 28
print counter
```

```
time.sleep(.25)
counter = counter * 2
print counter
counter = counter / 3
time.sleep(.25)
print counter
time.sleep(.25)
for counter in range(2,6):
  print counter
  time.sleep(.25)
for counter in range(10,4,-1):
  print counter
  time.sleep(.25)
while counter < 22:
  print counter +2
  counter = counter + 4
  time.sleep(.25)
list = ["a", "b", "c", "d", "e", "f", "g", "etc."]
print list
time.sleep(.25)
dictionary = {'Apple' : 'Fruit', 'Bush' : 'Plant', 'Carrot' : 'Vegetable'}
print dictionary
time.sleep(.25)
print 'Now we will add "dog" to the dictionary.'
time.sleep(.25)
dictionary['Dog'] = 'Animal'
print dictionary
time.sleep(.25)
del dictionary['Apple']
print 'Now as you can see, we have deleted "Apple" from the dictionary.'
print dictionary
```

and run your code!

Even if you entered the code correctly, you will have an error! Do you know why? We forgot to import the module called "time."

Add this to the very top of your code:

```
import time
```

and run your code again.

Congratulations! It works! In this code, we have used what we covered in this chapter.

End of Chapter Quiz

Can you answer the following questions?

1. What does the symbol "!" typically mean in Python?
2. What is a loop?
3. What's the difference between a list and a dictionary?
4. What is a function?

CHAPTER SIX

In this chapter we will cover:

- Modules
- Programs
- Two Python coding exercises

The product of this chapter is a student who is familiar with all the basics of Python.

Build Your Own

Create a program on your own that contains the following:

1.   Multiplies a number to the power of another   number

239925544. An if, elif, and else statement

239926160. A for loop

239926384. A while loop

239926552. A dictionary

Modules

Python has many modules that you can import into your project. These are pre-made sets of code that allow you to perform certain functions. Type and execute the following in the text editor:

```
print math.sqrt(64)
```

You received an error because the math module hasn't been imported yet. So, type the following above your code:

```
import math
```

This brings in additional math functions, such as finding the square root of something. Now your code should run!

To randomly choose a number, we can use the random integer function. Type the following in the text editor:

```
import random
```

```
print(random.randint(0,100))
```

and execute your code.

Create Modules

Type the following in the text editor:

```
import random
import math
from random import randint
def subtract_5(A):
    return A - 5
def add_10(A):
    return A + 10
def multiply_2(A):
    return A * 2
def random(A):
    B = randint(1,5)
    return A + B
```

Save the program as PythonModule.py. We have now created a module with four functions.

Open a new file and type the following within the text editor:

```
import PythonModule
print PythonModule.subtract_5(10)
print PythonModule.add_10(5)
print PythonModule.multiply_2(4)
print PythonModule.random(0)
```

and run the code. You just utilized the module you created and the functions therein! Review your code; can you tell what each line does?

Raw Input

The Raw Input function allows the user to enter information. We will now incorporate user input into our programs.

Enter the following in the text editor:

```
Date = raw_input('Please enter the date: ')
print 'The date you entered is ' + Date + '.'
```

and execute your code.

Let's try another raw input and use the if, elif, and else statements. Enter the following in the text editor:

```
print 'We are going to find out whether or not you like candy.'
Candy = raw_input('Do you like candy?: ')
if Candy == 'Yes':
    print 'You like candy!'
elif Candy == 'No':
    print 'You do not like candy...'
else:
    print 'Please enter Yes or No.'
```

Program

We will now create a program using Python. The program will store information about various people and then grab data (height, weight, birth year, etc.) by typing in their name. The first step will be to create a dictionary so we can easily handle the information. Open a New File and type the following in the text editor:

```
people_dictionary = {'Brett' : ['Male' , 'Weight 175'], 'Nancy' : ['Female' , 'Weight 125'], 'Patrick' : ['Male' , 'Weight 195'], 'Diane' : ['Female' , 'Weight 115'], 'Adam' : ['Male' , 'Weight 215']}
print people_dictionary
```

and execute your code.

Now we want to take information from the dictionary to show it to the user. To search for a person's name, type the following in the text editor:

```
Name = raw_input('Please type in a name: ')
print 'You typed in the name ' + Name
Persons_Data = people_dictionary[Name]
print Persons_Data
```

and run the code. If you receive an error, it means you didn't exactly type a name contained in the dictionary. Otherwise, it will run.

Run your program again and intentionally enter a name that isn't in the dictionary. Let's get rid of the error message and display a string that will explain to the user what happened and what needs to be done.

There are some other functions called "try" and "except." "Try" attempts to find things in a dictionary and, if found, does something. "Except" gives another action if something isn't found in the dictionary.

Edit your code so it is written as follows (we are adding try and except):

```
people_dictionary = {'Brett' : ['Male' , 'Weight 175'], 'Nancy' : ['Female' , 'Weight 125'], 'Patrick' : ['Male' , 'Weight 195'], 'Diane' : ['Female' , 'Weight 115'], 'Adam' : ['Male' , 'Weight 215']}
print people_dictionary
Name = raw_input('Please type in a name: ')
print 'You typed in the name ' + Name + '!'
try:
    Persons_Data = people_dictionary[Name]
    print Persons_Data
except:
    print 'That name, ' + Name + ', was not found in the dictionary.'
```

As you may have noticed, if you don't enter a name exactly as it's written in the dictionary, the program will think the name isn't there. For example, run your program and type brEtt or patRICK (as opposed to Brett or Patrick). To fix this, edit your code as follows (make all names lowercase and add .lower() and .capitalize() as below):

```
people_dictionary = {'brett' : ['Male' , 'Weight 175'],
    'nancy' : ['Female' , 'Weight 125'],
    'patrick' : ['Male' , 'Weight 195'],
    'diane' : ['Female' , 'Weight 115'],
    'adam' : ['Male' , 'Weight 215']}
print people_dictionary
Name = raw_input('Please type in a name: ').lower()
print 'You typed in the name ' + Name.capitalize()
try:
    Persons_Data = people_dictionary[Name]
    print Persons_Data
except:
    print 'That name (as written) was not found in the dictionary.'
```

Now run the code again and try entering one of the names in the dictionary in an odd format (like eMILy or something). It should work. This is called sanitizing your inputs. It helps minimize the likelihood of user error. Now we are going to better display the output. Edit the try function to the following:

```
try:
    Persons_Data = people_dictionary[Name]
    print 'Name: ' + Name.capitalize()
    print 'Sex: ' + Persons_Data[0]
    print 'Weight: ' + Persons_Data[1]
```

and run the program.

Now how do we allow people to search for more names or search again when they don't type the name correctly? First, we need to define a start point to our program (which will allow us to have people go to the start again). To do so, we add the following to the beginning of our code:

```
def start():
```

This will require you to then tab all of your code over once to the right (if it doesn't automatically tab). Now add the following to the end of your code:

```
start()
```

This will call the start function. Now let's add another step to the program that gives the user an option of searching for another name. At the very end of your program, right above start(), enter the following code:

```
def more():
    More = raw_input('Would you like to search for another name?: ')
    if More == 'No':
        quit()
    if More == 'Yes':
        start()
    else:
        print "Please enter Yes or No."
        more()
```

Note: quit() ends the program.

To direct the start() section of the program to the more() section of the program, enter more() at the end of the try function as follows:

```
try:
    Persons_Data = people_dictionary[Name]
    print 'Name: ' + Name.capitalize()
    print 'Sex: ' + Persons_Data[0]
    print 'Weight: ' + Persons_Data[1]
```

more()

And also enter more() at the end of the *except* function:

```
except:
    print 'That name (as written) was not found in the dictionary.'
    more()
```

Your final code should look like this:

```
def start():
    print 'Welcome!'
    people_dictionary = {'brett' : ['Male','Weight 175'],
        'nancy' : ['Female','Weight 125'],
        'patrick' : ['Male','Weight 195'],
        'diane' : ['Female','Weight 115'],
        'adam' : ['Male','Weight 215']}
    Name = raw_input('Please type in a name: ').lower()
    print 'You typed in the name ' + Name.capitalize()
    try:
        Persons_Data = people_dictionary[Name]
        print 'Name: ' + Name.capitalize()
        print 'Sex: ' + Persons_Data[0]
        print 'Weight: ' + Persons_Data[1]
        more()
    except:
        print 'That name (as written) was not found in the dictionary.'
        more()
def more():
    More = raw_input('Would you like to search for another name?: ')
    if More == 'No':
        return
    if More == 'Yes':
        start()
    else:
        print "Please enter Yes or No."
        more()

start()
```

Run the program. Now we have a fully operational program!

Python Exercise 1

Write a program in Python 2.7 using IDLE that demonstrates the following concepts. Most of this has been covered earlier, but some of it is new. On any point you don't understand, research online for the data and solution. Use comments in your program to denote where you demonstrate each step. If you cannot demonstrate any of these, research them online first. Here are your assigned tasks:

1. Assign an integer to a variable.

2. Assign a string to a variable.

3. Assign a float to a variable.

4. Use the print function to print out the variable you assigned.

5. Use each of these operators: +, *, /, =, %.

6. Use each of these logical operators: and, or, not.

7. Use each conditional statement: if, elif, else.

8. Use a *while* loop.

9. Use a *for* loop.

10. Create a list and iterate through that list using a *for* loop to print each item out on a new line.

11. Create a tuple and iterate through it using a *for* loop to print each item out on a new line.

12. Define a function that returns a string variable.

13. Call the function you defined above and print the result to the shell.

Note: It is okay, and even encouraged, to combine any of the above requirements. For example, you may use variables and operators within an *if else* statement, etc.

Python Exercise 2

To give you an idea of what you will run across as a developer, you will now be assigned a task that will require you to figure out a solution on your own. Here is your task:

Create a list of 10 items (such as animals, fruits, names, etc.). Ensure these items are strings values.

Now write a program that will alphabetize the list and then display the alphabetized list.

End of Chapter Quiz

Can you answer the following questions?

1. What is a module?
2. What is a Raw Input and how do you create one?
3. What should you do if you receive an error message when writing code, and you don't know what the error is?

CHAPTER SEVEN

In this chapter, you will be assigned exercises. Some of the tasks assigned to you will have been covered in the earlier parts of this class, and some will be new territory, requiring you to research online for solutions.

The key to programming is: problem-solving. As a programmer, you will run into many walls and have to solve many issues. The solution for this is searching for solutions online.

If you find yourself stuck on a particular point, attempt to find a solution on the Internet. Stack Overflow (https://stackoverflow.com) is a great resource.

The product of this chapter is a programmer who understands the basics of Python.

Python Exercise 1

Create a program that figures out how long you have been alive in terms of years and months.

Here is one way the code could look:

```
Name = raw_input ('Name: ')
print "Hello " + Name + "! We are going to find out how long you have been alive!"
Age = int(raw_input('How old are you?: '))
print "You are " + str(Age) + " years old."
Months = Age * 12
# 12 is the number of months in a year
Days = Age * 365
# 365 is the number of days in a year
print Name + " has been alive for about: " + str(Months) + " months and " + str(Days) + " days!"
```

Python Exercise 2

Modify the program you created in Python Exercise 1 to additionally display how many <u>minutes</u> and <u>seconds</u> are in the number of years that the user enters.

Hint: there are 525948 minutes in a year and 31556926 seconds in a year.

Python Exercise 3

The Python range() function generates a list of numbers, which is generally used to iterate over with *for* loops. In other words, using it creates the list of index numbers that you can then use in a *for* loop. The range() function has two different sets of parameters that can be used as follows:

range(stop): The number of numbers (integers) to generate, starting from zero.

range([start,] stop[, step]) //notice how the optional parameters are enclosed in [ ]. *start*: The starting number of the sequence. *stop*: The number of numbers (integers) to generate, starting from the integer specified with the start parameter. *step*: The difference between each number in the sequence. This is basically the 'count by' number.

Note that:

- All parameters must be integers.
- All parameters can be positive or negative.

The stop parameter is not the number the function will stop on but rather the Nth number produced, where stop is the Nth number.

For example, type and run the following code in a New Window:

```python
my_list = ['one', 'two', 'three', 'four', 'five']
my_list_len = len(my_list)
for i in range(0, my_list_len):
    print(my_list[i])
```

In this example, there are two parameters listed: 0 and my_list_len

Here are your tasks:

Step 1:

Open a New Window and use the Python range() function with one parameter to display the following:
0
1
2
3

Step 2:

Use the Python range() function with 3 parameters to display the following:
3
2
1
0

Step 3:

Use the Python range() function with 3 parameters to display the following:
8
6
4
2

Python Exercise 4

Watch this video in full and create a Python game (you can either copy the one in the video exactly or make your own):

https://www.youtube.com/watch?v=RliPr9V8f6A&t=1s

You Finished!

Congratulations on completing our Python book! If you are interested in delving deeper into coding, please contact The Tech Academy at learncodinganywhere.com

As the next step, we recommend enrolling in our Software Developer Boot Camp that will train you as a well-rounded, entry-level software developer!

INDEX

Made in the USA
Columbia, SC
29 June 2019